SOCIAL SCIENCE

Daredevils
Who Conquered the Skies

MICHÈLE DUFRESNE

TABLE OF CONTENTS

PIONEER VALLEY EDUCATIONAL PRESS, INC

CLOSE TO THE SUN

Since the beginning of human existence, people have dreamed of flying. The ancient Greek myth of Icarus (*Ick-ah-riss*) tells the tale of a man who flew using wax wings and feathers. Although his father warned him not to fly near the sun, Icarus ignored him, and his wings melted, causing him to fall into the sea and die. This story is the **origin** of the phrase *flying too close to the sun*, which means you are overconfident and reckless.

People have attempted many ways to fly like a bird, some of them dangerous. Over time, inventors tried many methods that did not succeed, but with every failure came new ideas. What was once a dream became a reality thanks to so many brave men and women.

MORE TO EXPLORE

When aviation pioneer **JAMES DOOLITTLE** was only 15 years old, he built a glider, which he tried to fly by jumping off a cliff. He crashed, but later he became an ace pilot and set many records.

GETTING OFF THE GROUND

The earliest flying machines were giant bags of fabric filled with heated air. The first of these manned balloons flew over a city in France and quickly became all the rage. Balloons were exciting because people got to experience being up high. But this wasn't really flying; balloons were guided by a rope or just floated wherever the wind took them.

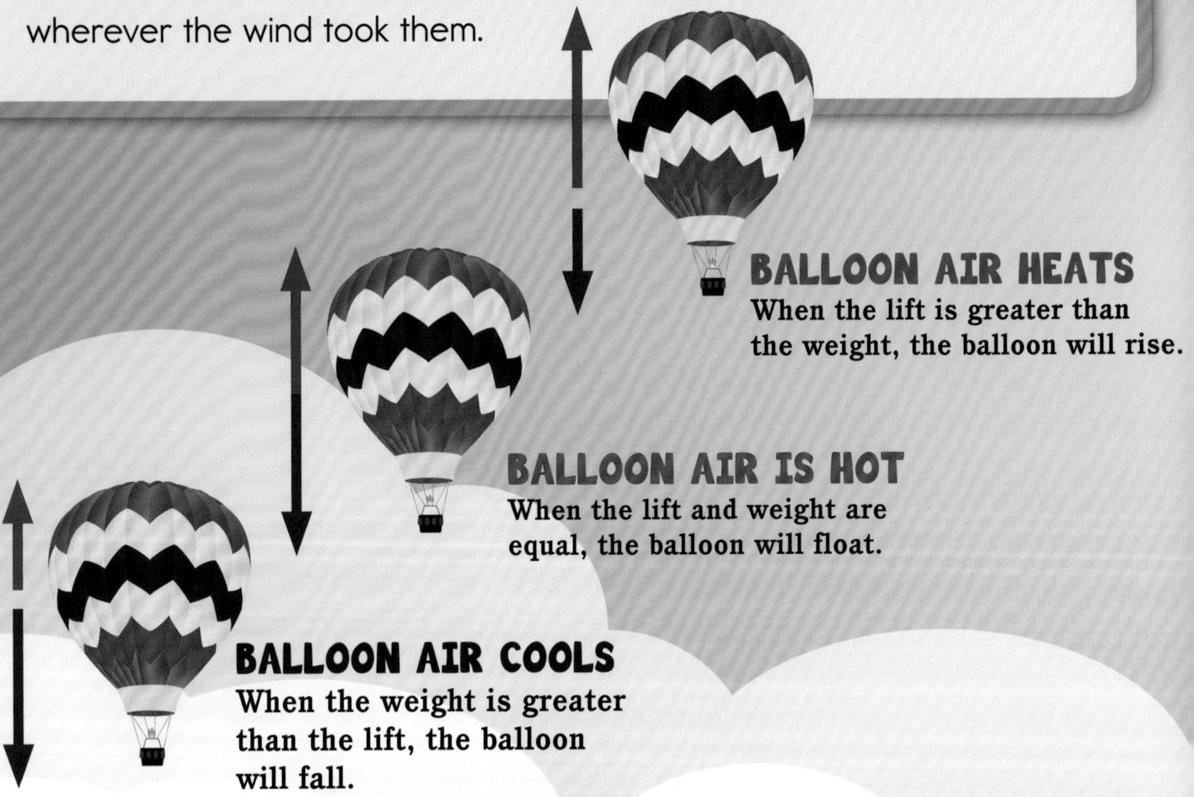

BALLOON AIR HEATS
When the lift is greater than the weight, the balloon will rise.

BALLOON AIR IS HOT
When the lift and weight are equal, the balloon will float.

BALLOON AIR COOLS
When the weight is greater than the lift, the balloon will fall.

The first hot-air balloon passengers weren't people. Instead, the honor went to a sheep, rooster, and duck.

Engineers began to experiment with **gliders**, which were heavier than air and had no engines. German engineer Otto Lilienthal built a 50-foot hill near his home and tried out various-shaped wings. He waited for the right wind conditions, then ran down his hill, soaring into the air. He became known as "Birdman" and "Flying Man." In 1896, when experimenting with a new design, he fell. Mortally wounded, his famous last words were, "Sacrifices must be made."

Otto Lilienthal testing one of his gliders

In 1903, Wilbur and Orville Wright built the *Wright Flyer*, the first powered airplane. It flew for 59 seconds and reached a height of 852 feet. The brothers continued to work on their machine, making improvements until it could fly for miles and miles, climb hundreds of feet off the ground, and complete amazing airborne turns, including figure eights.

MORE TO EXPLORE

After December 17, 1903, the *Wright Flyer* never went airborne again. The plane was damaged beyond repair after a **STRONG WIND GUST** flipped it over multiple times.

Wilbur went to Europe, where he gave rides to **journalists** and government leaders. The Wright brothers became wealthy businessmen by selling their airplanes, first in Europe and later in the United States.

The original *Wright Flyer* is on exhibit at the National Air and Space Museum in Washington, DC.

WAR IN THE SKY

Airplanes were invented just 11 years before the First World War. At first, planes were mostly used for **reconnaissance**. Pilots would fly over a battlefield to study the enemy's movements. Later, the military added small bombs for aircraft to drop onto enemy locations. This was dangerous, since planes needed to be close to the ground to drop bombs, and this left them vulnerable to attack. Eventually, bombs were developed that could be delivered from higher elevations.

Planes were also equipped with machine guns, and pilots began to battle each other in the sky. These were called dogfights: clashes in the air between fighter planes at close range.

The first airplanes had no military markings. This caused confusion, and soldiers mistakenly shot down planes from their own side. Finally, each nation began placing markings on the sides, wings, and fins or rudders of a plane to help identify it from the ground and in the air.

The best pilots, ones who brought down more than five airplanes, were called **ACES**. The Red Baron, a famous German pilot, was well known for shooting down 80 planes.

FLYING FIRSTS

In 1913, a British newspaper offered a prize of 10,000 pounds sterling—more than a million dollars today—to the first **aviator** to fly across the Atlantic Ocean. After World War I ended in 1918, so many teams were vying for the prize that it was difficult to find a runway to use.

British aviators John Alcock and Arthur Brown flew through rain and snow from Newfoundland to Ireland. At one point, they spiraled 4,000 feet down until coming just 100 feet short of crashing into the ocean. They were able to level out the plane and put it down in a **bog**. John and Arthur flew about 1,890 miles in 16 hours and were awarded the prize money. The British king knighted the two men for their accomplishment.

On May 21, 1927, Charles Lindbergh became the first solo pilot to fly nonstop across the Atlantic Ocean from New York to Paris, France. This amazing event enthralled the world, and many waited near their radios, listening for news of the historic flight. More than 150,000 people greeted Charles in Paris and cheered as he landed.

MORE TO EXPLORE

One of Charles's biggest challenges on his 33½-hour flight was **STAYING AWAKE**. He flew very close to the ocean, hoping that the chilly spray would help combat his exhaustion.

Bessie Coleman, like many, was fascinated with flying after hearing stories from World War I veterans. Bessie decided to learn how to fly, but because she was a young African American woman, no flight school would accept her. Undaunted, she traveled to France, where she got her pilot's license in 1921. Returning to the United States, she became a star, participating in air shows around the country. Bessie died in an airplane crash in 1926, but she is remembered for breaking down barriers as one of the earliest female aviators and the first African American woman to receive a pilot's license.

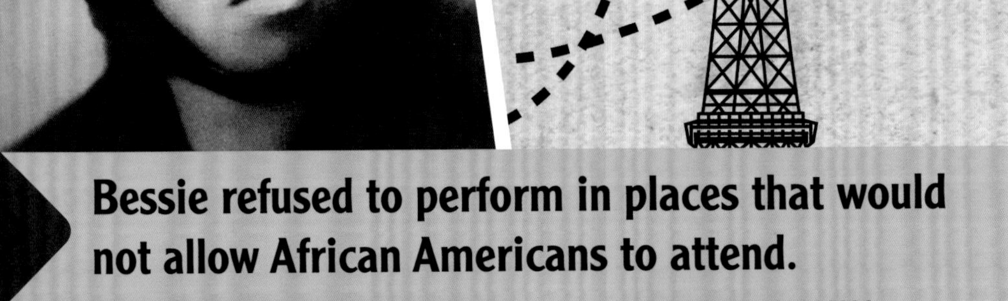

Bessie refused to perform in places that would not allow African Americans to attend.

Amelia Earhart set many aviation records and became the first woman to fly solo across the Atlantic Ocean in 1932. She served as a Red Cross nurse's aide during World War I and worked near a local airfield, where she watched pilots training. After the war, she took flying lessons and purchased a yellow plane she nicknamed "the Canary."

In 1937, Amelia and her **navigator** attempted to fly around the world. With only 7,000 of their 29,000-mile journey remaining, the pair lost radio contact and were never seen or heard from again. Some believe their plane ran out of fuel and fell into the ocean, but others think they crash-landed on an uninhabited island. Some **artifacts** have been found on the island of Nikumaroro, supporting the theory that the duo may have ended up there.

INTO SPACE

Over the years, humans' ability to fly led to dreams of traveling beyond the boundaries of Earth. Their next goal was to explore even greater mysteries in the far reaches of space.

At the end of World War II, the United States and the Soviet Union entered the Cold War, a period when the two powerful nations disagreed over ideas such as people's freedoms. The two countries began to compete in the Space Race. In 1957, the Soviets launched Sputnik 1, the first man-made **satellite**. Americans were concerned that the Soviets appeared to be winning the Space Race.

Before humans began traveling into space, the two countries sent animals such as dogs, monkeys, and a rabbit into space as passengers. NASA wanted to know what it might be like for an astronaut in space, so a group of chimpanzees were trained. Ultimately, one named Ham was selected to go up. During the flight, he was tested on how he reacted to **acceleration** and being weightless, as well as his ability to perform some simple tasks. Ham became instantly famous.

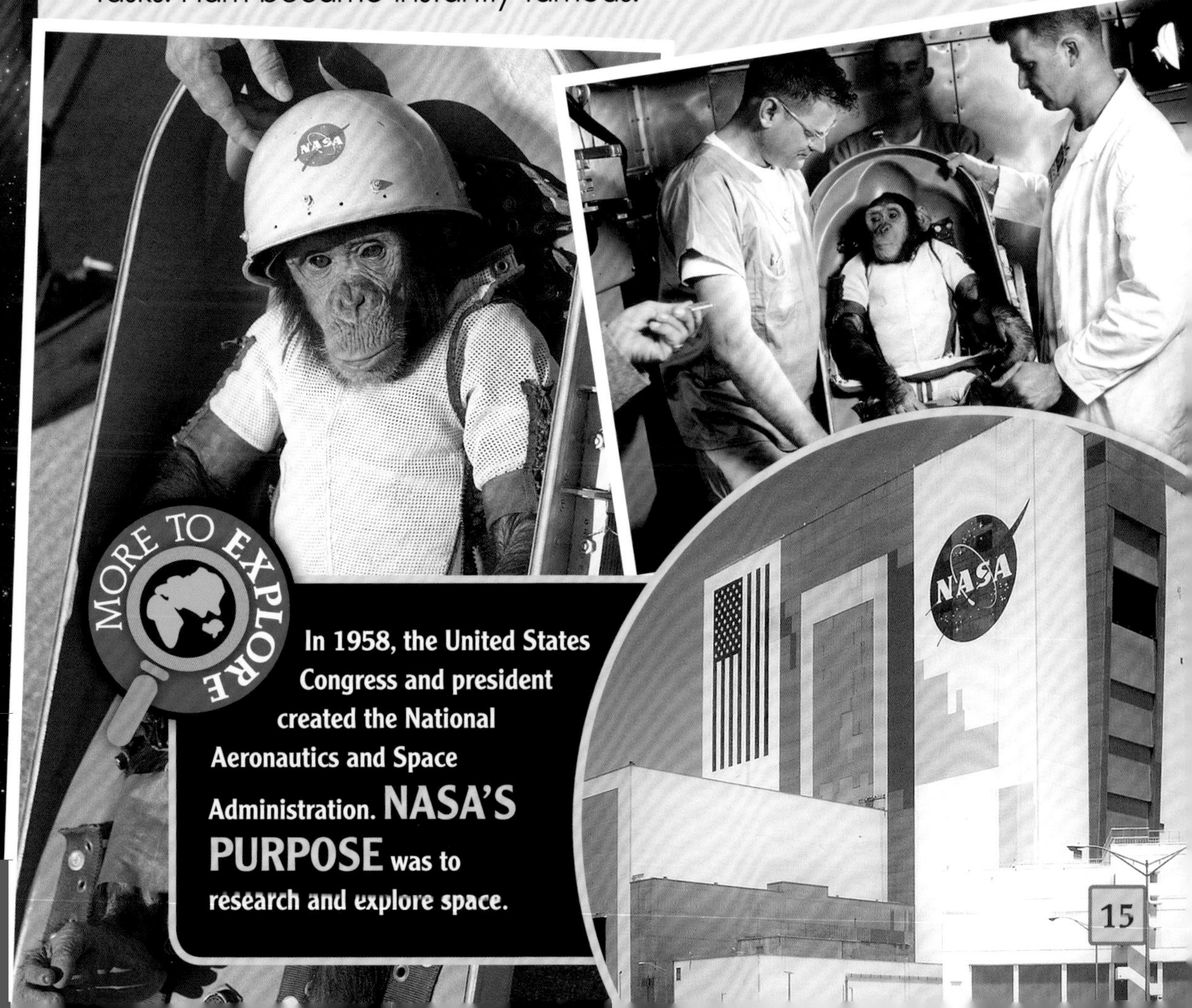

MORE TO EXPLORE

In 1958, the United States Congress and president created the National Aeronautics and Space Administration. **NASA'S PURPOSE** was to research and explore space.

The world was taken by surprise when Yuri Gagarin, a Soviet **cosmonaut**, launched into space on April 12, 1961, and became the first person in space. The manned flight was short, lasting less than two hours as he orbited Earth. The United States quickly sent up its own astronaut. Alan Shepard flew in May, rising 116 miles into space during a 15½-minute flight.

Yuri Gagarin

In 1969, *Apollo 11* blasted into space carrying three astronauts: Neil Armstrong, Edwin "Buzz" Aldrin, and Michael Collins. Four days later, Neil and Buzz landed the *Eagle*, a small lunar craft, on the moon. On July 20, Neil became the first man to step onto the moon. He and Buzz also took photos, did some experiments, and picked up rocks and dirt to bring back to Earth. They planted an American flag in the ground.

On his historic mission, Neil carried pieces of the original *Wright Flyer* in his space-suit pocket.

In 1963, a Soviet cosmonaut named Valentina Tereshkova was the first woman to fly into space. Sally Ride became the first American woman on June 18, 1983, when she flew on the shuttle *Challenger*. During the trip, astronauts launched two communication satellites, and Sally became the first woman to operate the shuttle's robotic arm in space.

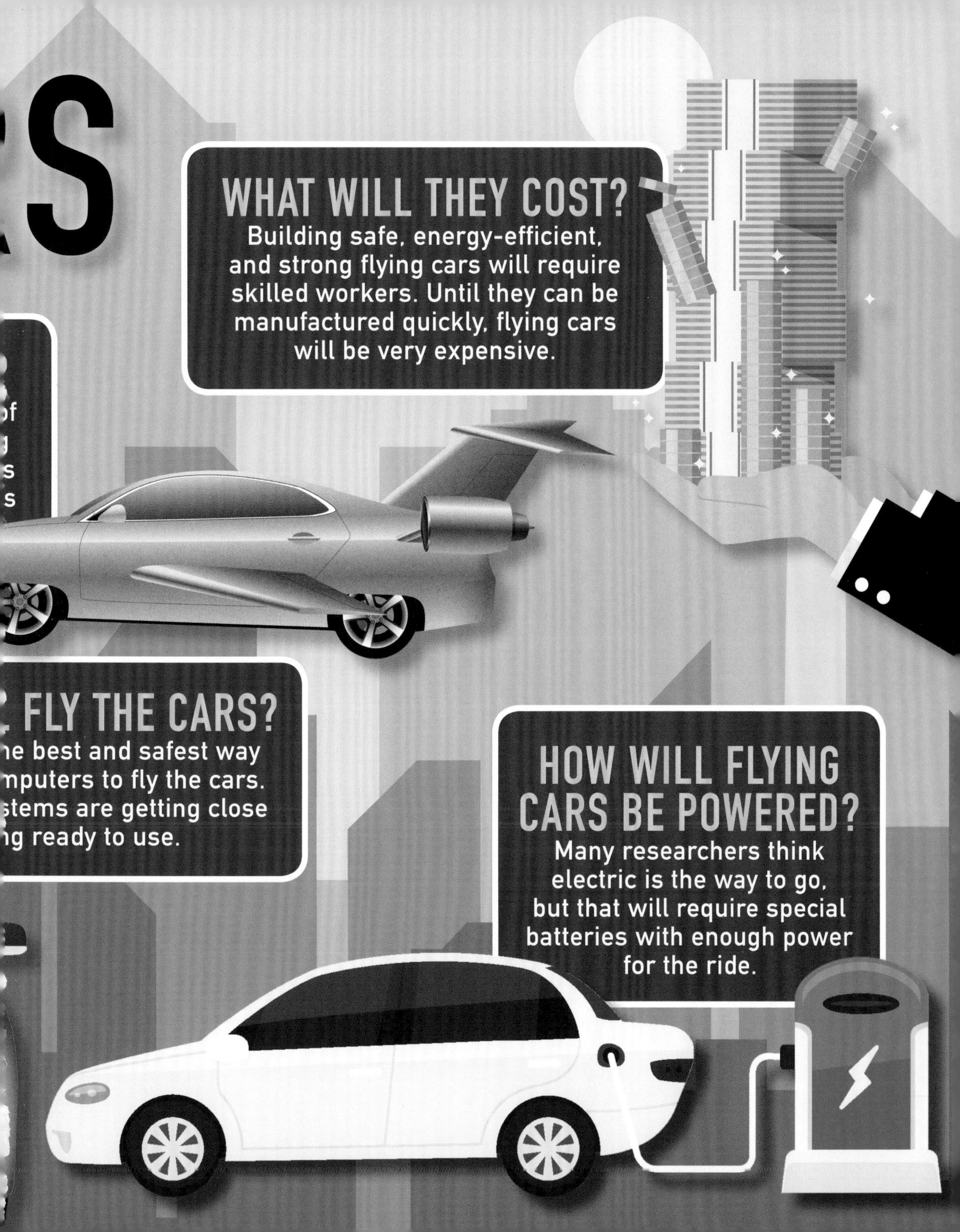

WHAT WILL THEY COST?
Building safe, energy-efficient, and strong flying cars will require skilled workers. Until they can be manufactured quickly, flying cars will be very expensive.
FLY THE CARS?
ne best and safest way
nputers to fly the cars.
stems are getting close
ng ready to use.
HOW WILL FLYING CARS BE POWERED?
Many researchers think electric is the way to go, but that will require special batteries with enough power for the ride.

Today, people continue to explore the sky, hoping to learn more about what lies in deep space. NASA is developing more advanced rockets and spacecraft that will soon lead us into the outer reaches of our universe.

Humans have come so far since the early days of flight, when inventors gambled on little more than cloth and fragile frames to keep them airborne. It will be amazing to see what the world learns as future technologies let people fly higher, farther, and faster than ever before.

GLOSSARY

acceleration
the rate at which the speed of something changes over time

artifacts
old, found objects

aviator
a person who flies airplanes or other flying machines

bog
soft, wet ground

cosmonaut
a Soviet or Russian astronaut

gliders
aircraft with wings but no motor

journalists
people who write news for newspapers, magazines, and other media outlets

navigator
a person who figures out how to get to a place

origin
beginning

reconnaissance
military soldiers or airplanes sent to gain information about the enemy

satellite
a machine orbiting in space for communicating or recording information

INDEX

FLYING CAR

Wouldn't it be great to avoid traffic and potholes on the ground and simply fly to school or work? Today, companies are working on building the first flying cars. What are some of the obstacles to getting this new technology?

WHAT WILL THEY SOUND LIKE?

Helicopters and airplanes make a great deal noise. How would we feel about the sky bein filled with even more noisy aircraft? Engineer will need to make a quiet flying car that blenc into the sounds around us.

P

WHERE WILL PEOPLE LAND?

Special airports will have to be built, especially in urban areas.

WHO WIL

Most likely, t will be for co Self-flying sy to bei